ENERGY SECTOR STANDARD OF THE PEOPLE'S REPUBLIC OF CHINA

中华人民共和国能源行业标准

Code for Design of Concrete Face Rockfill Dams

混凝土面板堆石坝设计规范

NB/T 10871-2021

Replace DL/T 5016-2011

Chief Development Department: China Renewable Energy Engineering Institute

Approval Department: National Energy Administration of the People's Republic of China

Implementation Date: June 22, 2022

China Water & Power Press

中国水利水电出版社

Beijing 2022

All rights reserved. No part of this publication may be reproduced, stored in a retrieval system, or transmitted in any form or by any means—electronic, mechanical, photocopying, recording or otherwise, without prior written permission of the publisher.

图书在版编目（ＣＩＰ）数据

混凝土面板堆石坝设计规范：NB/T 10871-2021 = Code for Design of Concrete Face Rockfill Dams (NB/T 10871-2021)：英文 / 国家能源局发布. -- 北京：中国水利水电出版社，2022.8
ISBN 978-7-5226-0976-8

Ⅰ．①混… Ⅱ．①国… Ⅲ．①混凝土面板坝－堆石坝－设计规范－中国－英文 Ⅳ．①TV641.4-65

中国版本图书馆CIP数据核字(2022)第164231号

ENERGY SECTOR STANDARD
OF THE PEOPLE'S REPUBLIC OF CHINA
中华人民共和国能源行业标准

Code for Design of Concrete Face Rockfill Dams
混凝土面板堆石坝设计规范
NB/T 10871-2021
Replace DL/T 5016-2011
（英文版）

First published 2022
Issued by National Energy Administration of the People's Republic of China
国家能源局　发布
Translation organized by China Renewable Energy Engineering Institute
水电水利规划设计总院　组织翻译
Published by China Water & Power Press
中国水利水电出版社　出版发行

　　Tel: (+ 86 10) 68545810　68545874
　　leezhe@mwr.gov.cn
　　Account name: China Water & Power Press
　　Account number: 0200096319000089691
　　Address: No.1, Yuyuantan Nanlu, Haidian District, Beijing 100038, China
　　Organization code: 400014639
　　http://www.waterpub.com.cn

中国水利水电出版社微机排版中心
北京中献拓方科技发展有限公司
184mm×260mm　16开本　4.25 印张　135 千字
2022 年 8 月第 1 版　2022 年 8 月第 1 次印刷

Price（定价）：￥720.00 (US $ 102.00)

Introduction

This English version is one of China's energy sector standard series in English. Its translation was organized by China Renewable Energy Engineering Institute authorized by National Energy Administration of the People's Republic of China in compliance with relevant procedures and stipulations. This English version was issued by National Energy Administration of the People's Republic of China in Announcement [2022] No. 4 dated May 13, 2022.

This version was translated from the Chinese Standard NB/T 10871-2021, *Code for Design of Concrete Face Rockfill Dams*, published by China Water & Power Press. The copyright is reserved by National Energy Administration of the People's Republic of China. In the event of any discrepancy in the implementation, the Chinese version shall prevail.

Many thanks go to the staff from the relevant standard development organizations and those who have provided generous assistance in the translation and review process.

For further improvement of the English version, any comments and suggestions are welcome and should be addressed to:

China Renewable Energy Engineering Institute

No. 2 Beixiaojie, Liupukang, Xicheng District, Beijing 100120, China

Website: www.creei.cn

Translating organization:

POWERCHINA Kunming Engineering Corporation Limited

China Renewable Energy Engineering Institute

Translating staff:

| FENG Yelin | HUANG Qingfu | JIA Haibo | MIAO Jiali |
| LI Bowen | KONG Lingxue | WANG Fuqiang | LIU Rongli |

Review panel members:

JIN Feng	Tsinghua University
ZHANG Zongliang	Member of Chinese Academy of Engineering
	POWERCHINA Kunming Engineering Corporation Limited
LIU Xiaofen	POWERCHINA Zhongnan Engineering Corporation

	Limited
LI Zhongjie	POWERCHINA Northwest Engineering Corporation Limited
GUO Jie	POWERCHINA Beijing Engineering Corporation Limited
XU Zeping	China Institute of Water Resources and Hydropower Research
QIE Chunsheng	Senior English Translator
QIAO Peng	POWERCHINA Northwest Engineering Corporation Limited
PENG Fuping	POWERCHINA Kunming Engineering Corporation Limited
QI Wen	POWERCHINA Beijing Engineering Corporation Limited
HAO Peng	POWERCHINA Guiyang Engineering Corporation Limited
YANG Xing	POWERCHINA Chengdu Engineering Corporation Limited
WANG Yingjun	POWERCHINA Huadong Engineering Corporation Limited
CHEN Lei	POWERCHINA Zhongnan Engineering Corporation Limited
ZHANG Ming	Tsinghua University
CHENG Jing	Hohai University
HOU Yujing	China Institute of Water Resources and Hydropower Research
YAN Wenjun	Army Academy of Armored Forces, PLA
YOU Yang	China Society for Hydropower Engineering
LI Shisheng	China Renewable Energy Engineering Institute

National Energy Administration of the People's Republic of China

翻译出版说明

本译本为国家能源局委托水电水利规划设计总院按照有关程序和规定，统一组织翻译的能源行业标准英文版系列译本之一。2022年5月13日，国家能源局以2022年第4号公告予以公布。

本译本是根据中国水利水电出版社出版的《混凝土面板堆石坝设计规范》NB/T 10871—2021翻译的，著作权归国家能源局所有。在使用过程中，如出现异议，以中文版为准。

本译本在翻译和审核过程中，本标准编制单位及编制组有关成员给予了积极协助。

为不断提高本译本的质量，欢迎使用者提出意见和建议，并反馈给水电水利规划设计总院。

地址：北京市西城区六铺炕北小街2号
邮编：100120
网址：www.creei.cn

本译本翻译单位： 中国电建集团昆明勘测设计研究院有限公司
　　　　　　　　　水电水利规划设计总院

本译本翻译人员： 冯业林　黄青富　贾海波　缪嘉莉
　　　　　　　　　李博文　孔令学　王富强　刘荣丽

本译本审核人员：

　　金　峰　清华大学

　　张宗亮　中国工程院院士，中国电建集团昆明勘测设计研究院有限公司

　　刘小芬　中国电建集团中南勘测设计研究院有限公司

　　李仲杰　中国电建集团西北勘测设计研究院有限公司

　　郭　洁　中国电建集团北京勘测设计研究院有限公司

　　徐泽平　中国水利水电科学研究院

　　郄春生　高级翻译

　　乔　鹏　中国电建集团西北勘测设计研究院有限公司

　　彭富平　中国电建集团昆明勘测设计研究院有限公司

齐　文	中国电建集团北京勘测设计研究院有限公司
郝　鹏	中国电建集团贵阳勘测设计研究院有限公司
杨　星	中国电建集团成都勘测设计研究院有限公司
王樱畯	中国电建集团华东勘测设计研究院有限公司
陈　蕾	中国电建集团中南勘测设计研究院有限公司
张　明	清华大学
程　井	河海大学
侯瑜京	中国水利水电科学研究院
闫文军	中国人民解放军陆军装甲兵学院
由　洋	中国水力发电工程学会
李仕胜	水电水利规划设计总院

国家能源局

Announcement of National Energy Administration of the People's Republic of China [2021] No. 6

In accordance with *Standardization Law of The People's Republic of China* and *Measures for the Administration of Energy Sector Standardization*, National Energy Administration of the People's Republic of China has approved and issued 356 energy sector standards including *Code for Engineering Design of Underground Forced Permeability Enhancement for Pre-drainage of Coal Gas* (Attachment 1) and the foreign language version of 25 energy sector standards including *Technical Code for Design and Calculation of Combustion System of Fossil-fired Power Plant* (Attachment 2).

Attachments: 1. Directory of Sector Standards

2. Directory of Foreign Language Version of Sector Standards

National Energy Administration of the People's Republic of China

December 22, 2021

Attachment 1:

Directory of Sector Standards

Serial number	Standard No.	Title	Replaced standard No.	Adopted international standard No.	Approval date	Implementation date
...						
22	NB/T 10871-2021	Code for Design of Concrete Face Rockfill Dams	DL/T 5016-2011		2021-12-22	2022-06-22
...						

Foreword

According to the requirements of Document GNZTKJ [2017] No. 52 issued by National Energy Administration of the People's Republic of China, "Notice on Releasing the Development and Revision Plan and the Translation and Publication Plan of the English Version of Energy Sector Standards in 2017", and after extensive investigation and research, summarization of practical experience, and wide solicitation of opinions, the drafting group has prepared this code.

The main technical contents of this code include: general provisions, terms and symbols, basic requirements, project layout and dam zoning, filling materials and filling criteria, plinth, face slab, joints and waterstops, foundation treatment, analysis and calculation, seismic measures, staged construction and dam heightening, safety monitoring, construction requirements, initial impoundment, operation and maintenance.

The main technical contents revised in this code are as follows:

— Adding terms "dam zoning" and "horizontal joint".

— Adding the requirements for extra-high dams with a height of 200 m or more.

— Adding the design requirements of slope fixing upstream of cushion material by concrete extruded curb or turnover formworks mortar.

— Adding the design requirements of toe wall.

— Adding the chapter "Construction Requirements".

— Adding the chapter "Initial Impoundment, Operation and Maintenance".

— Revising the scope of application and dam height classification.

— Revising the requirement for filling material gradation and filling.

— Revising the structural design requirements for face slab jointing, reinforcement, etc.

— Revising the requirement of joint waterstop structure and material

— Revising the stress and deformation calculation requirements and methods for dam and face slab.

National Energy Administration of the People's Republic of China is in charge of the administration of this code. China Renewable Energy Engineering Institute has proposed this code and is responsible for its routine management.

Energy Sector Standardization Technical Committee on Hydropower Investigation and Design is responsible for the explanation of specific technical contents. Comments and suggestions in the implementation of this code should be addressed to:

China Renewable Energy Engineering Institute
No. 2 Beixiaojie, Liupukang, Xicheng District, Beijing 100120, China

Chief development organizations:

POWERCHINA Kunming Engineering Corporation Limited
China Renewable Energy Engineering Institute

Participating development organizations:

POWERCHINA Northwest Engineering Corporation Limited

POWERCHINA Chengdu Engineering Corporation Limited

POWERCHINA Huadong Engineering Corporation Limited

POWERCHINA Zhongnan Engineering Corporation Limited

POWERCHINA Guiyang Engineering Corporation Limited

POWERCHINA Beijing Engineering Corporation Limited

China Three Gorges Corporation

Tsinghua University

Wuhan University

Nanjing Hydraulic Research Institute

Huaneng Lancang River Hydropower Inc.

CHN Energy Dadu River Hydropower Development Co., Ltd.

Chief drafting staff:

ZHANG Zongliang	FENG Yelin	ZHOU Heng	WANG Fuqiang
HUANG Wei	YU Xueming	WANG Yiming	ZHANG Bingyin
WANG Qingxiang	ZHAN Zhenggang	SUN Yi	MI Zhankuan
ZHOU Wei	AI Yongping	ZHU Yongguo	HUANG Qingfu
ZHANG Libing	ZHAO Lin	WANG Junli	DOU Xiangxian
QIU Huanfeng	WANG Xiaoliang	WU Jicai	ZHU Aili
YU Yuzhen	MA Gang	CHI Fudong	WEI Kuangmin

KONG Lingxue	YAN Shanglong	LU Xi	CAI Xinhe
LU Yuping	WU Weiwei	WU Yongkang	CAO Xuexing
MIAO Zhe	LI Hongxin	TANG Ke	

Review panel members:

YANG Zeyan	WU Gaojian	PAN Jiangyang	YAO Shuanxi
DENG Yiguo	LI Yonghong	LUO Guangqi	WANG Yuanliang
CHEN Zhenwen	XIONG Zebin	ZHANG Dongsheng	LI Guoying
ZHU Sheng	ZHANG Jianhai	LIU Shaochuan	LIU Chao
YIN Ai	LI Shisheng		

Contents

1	**General Provisions**	1
2	**Terms and Symbols**	2
2.1	Terms	2
2.2	Symbols	4
3	**Basic Requirements**	6
4	**Project Layout and Dam Zoning**	7
4.1	Project Layout	7
4.2	Dam Crest	7
4.3	Dam Slope	8
4.4	Dam Zoning	9
5	**Filling Materials and Filling Criteria**	12
5.1	Filling Materials Investigation, Testing and Quarry Planning	12
5.2	Cushion Materials and Transition Materials	13
5.3	Rockfill Materials	14
5.4	Filling Criteria	15
6	**Plinth**	17
6.1	Alignment and Layout of Plinth	17
6.2	Plinth Size	17
6.3	Toe Wall	18
6.4	Concrete and Reinforcement of Plinth and Toe Wall	19
7	**Face Slab**	21
7.1	Face Slab Jointing	21
7.2	Face Slab Thickness	21
7.3	Face Slab Concrete	22
7.4	Reinforcement Arrangement	23
7.5	Crack Control Measures for Face Slab	24
8	**Joints and Waterstops**	25
8.1	General Requirements	25
8.2	Perimeter Joints	25
8.3	Vertical Joints	27
8.4	Other Joints	29
8.5	Joint Filling and Waterstop Materials	30
9	**Foundation Treatment**	32
9.1	Excavation of Foundation and Abutments	32
9.2	Treatment of Foundation Defects	32
10	**Analysis and Calculation**	36
10.1	Seepage Analysis	36

10.2	Sliding Stability Analysis	37
10.3	Stress and Deformation Analysis	37
11	**Seismic Measures**	**39**
12	**Staged Construction and Dam Heightening**	**41**
12.1	Staged Construction	41
12.2	Temporary Water Retaining	41
12.3	Embankment Protection Under Overflow Condition	42
12.4	Dam Heightening	42
13	**Safety Monitoring**	**44**
14	**Construction Requirements**	**47**
15	**Initial Impoundment, Operation and Maintenance**	**49**
Explanation of Wording in This Code		**51**
List of Quoted Standards		**52**

1 General Provisions

1.0.1 This code is formulated with a view to standardizing the design of concrete face rockfill dams to achieve the objectives of safety and reliability, economic rationality, technological advancement, environment friendliness, and resource conservation.

1.0.2 This code is applicable to the design of concrete face rockfill dams for the construction, renovation, and extension of hydropower projects.

1.0.3 In addition to this code, the design of concrete face rockfill dams shall comply with other current relevant standards of China.

2 Terms and Symbols

2.1 Terms

2.1.1 concrete face slab

reinforced concrete impervious element on the upstream face of embankment

2.1.2 concrete face rockfill dam (CFRD)

roller compacted rockfill dam that employs a concrete face slab as the upstream impervious element

2.1.3 dam height

height measured from the lowest point of the plinth foundation surface, or from the lowest point of the foundation surface along the dam axis, to the dam crest, excluding the camber for settlement, trenches, and cavities backfill, whichever is greater

2.1.4 rockfill embankment

dam body filled with properly graded materials in zones downstream of the concrete face slab

2.1.5 dam zoning

division of dam body into different zones according to functions, filling materials, construction schedule, etc.

2.1.6 cushion zone

direct support of the face slab, which evenly transmits water pressure to the rockfill and serves as seepage control

2.1.7 fine cushion zone

cushion zone downstream of perimeter joint, serving as a filter for sealing materials of perimeter joint and adjacent joints of face slab

2.1.8 transition zone

zone between cushion zone and rockfill zone, serving as protection and transition

2.1.9 drainage zone

chimney and horizontal drain, constructed of highly pervious rockfill of rocks or gravels, in an embankment of sand and gravel or soft rock

2.1.10 upstream rockfill zone

upstream part of rockfill embankment, which acts as a major zone for bearing the water pressure

2.1.11 downstream rockfill zone

downstream part of rockfill embankment, which maintains the embankment stability along with the upstream rockfill zone

2.1.12 downstream slope protection

dry boulder masonry or other structures that are placed on the downstream slope face of the dam for enhancing the integrity and stability of the slope surface

2.1.13 upstream blanket zone

zone constructed of silt or similar materials, which covers the face slab, plinth and perimeter joint, and acts as secondary seepage control for the embankment

2.1.14 weighted cover zone

random rocks dumped on the upstream blanket zone to maintain a stable blanket

2.1.15 plinth

concrete slab connecting the foundation impervious element and the face slab, which can be flat, narrow, or inclined

2.1.16 plinth line

intersection line between the extension of face slab bottom surface and the design foundation surface of plinth, i.e. X line

2.1.17 toe wall

concrete base or retaining wall arranged along the plinth line and connected with face slab

2.1.18 concrete cutoff wall

underground diaphragm designed for seepage control in loose pervious foundation, dam, or cofferdam, which is constructed by placing concrete in excavated grooves or drilled secant pile holes whose walls have been stabilized by slurry

2.1.19 plinth extension

concrete or shotcrete plate on the dam foundation surface downstream or upstream of the plinth, used to extend the seepage path and reduce the foundation hydraulic gradient

2.1.20 concrete connection slab

concrete structure between the plinth and the cutoff wall in the foundation to adapt to the foundation deformation in cases where the plinth is built on the overburden

2.1.21 parapet wall

solid wall on the dam crest, which is connected to the top of face slab and is designed to prevent overtopping caused by wave runup

2.1.22 perimeter joint

joint between face slab and plinth or toe wall

2.1.23 vertical joint

joint between strips of the face slab. In general, the vertical joints of the tensile zone of face slab near bank slopes are referred to as tensile vertical joints, whereas the vertical joints of the compressive zone of face slab on the riverbed are referred to as compressive vertical joints. To prevent the extrusion damage of the face slab, the compressive vertical joints are filled with a certain width of compressible materials and are referred to as flexible vertical joints

2.1.24 horizontal joint

permanent horizontal structural joint set in the face slab at a certain elevation to improve the stress condition of the face slab

2.1.25 flexible filler

flexible material used as waterstop, which is composed of bitumen, rubber, and fillers

2.1.26 modulus-increased zone

special rockfill zone with a compression modulus higher than that in adjacent zones

2.2 Symbols

1A	upstream blanket zone
1B	weighted cover zone
2A	cushion zone
2B	fine cushion zone
3A	transition zone
3B	upstream rockfill zone

3C   downstream rockfill zone

3D   drainage zone

F   concrete face slab

T   plinth

P   downstream slope protection

α angle of the interface between upstream and downstream rockfill zones with respect to the plumb line

3 Basic Requirements

3.0.1 CFRDs shall be classified by height into low, medium, and high dams according to the following criteria:

1   low dams, when the height is below 30 m.

2   medium dams, when the height is between 30 m and 70 m.

3   high dams, when the height is over 70 m. A dam with a height of 200 m or more is termed an extra-high dam.

3.0.2 The grade of CFRD shall be determined in accordance with the current standards of China GB 50201, *Standard for Flood Control*; and DL 5180, *Classification and Design Safety Standard of Hydropower Projects*.

3.0.3 The design conditions or design situations of CFRD shall be in accordance with the current sector standard NB/T 10872, *Code for Design of Roller-Compacted Embankment Dams*.

3.0.4 CFRD shall meet the requirements for stability, seepage, deformation and freeboard for various load combinations under usual and unusual service conditions.

3.0.5 The rational service life of CFRD shall be in accordance with the current standards of China GB 50199, *Unified Standard for Reliability Design of Hydraulic Engineering Structures*; and NB/T 10857, *Code for Design of Rational Service Life and Durability for Hydropower Projects*.

3.0.6 Seismic design of CFRD shall comply with the current sector standard NB 35047, *Code for Seismic Design of Hydraulic Structures of Hydropower Project*.

3.0.7 In cases where the construction conditions are particularly complex or the design of CFRD is beyond the current technology level and the engineering experiece, special studies shall be conducted.

4 Project Layout and Dam Zoning

4.1 Project Layout

4.1.1 The CFRD project layout shall be determined through techno-economic comparison according to the topographical and geological conditions at the dam site to facilitate the arrangement of plinth and other structures as well as the construction.

4.1.2 The dam axis should be straight. In special cases, a broken or curved line may be adopted.

4.1.3 The rockfill dam may be constructed on dense riverbed overburden. When the overburden contains soft interlayers such as silty sand and clayey soil, its safety and economic rationality shall be evaluated through static and dynamic stability and deformation analysis of the embankment and overburden as well as dam foundation treatment.

4.1.4 The layout of water release structures and water conveyance structures shall comply with the current sector standard NB/T 10872, *Code for Design of Roller-Compacted Embankment Dams*.

4.1.5 CFRD should be provided with facilities to empty the reservoir or structures to lower the reservoir level. For dams that have a design seismic intensity of Ⅷ and above or Grade 1 and Grade 2 dams, emptying facilities shall be set for lowering the reservoir level. Further, emptying facilities may be integrated with structures for flood discharge, sediment flushing, water conveyance, river diversion, ecological flow, etc. If the emptying capability fails to meet the requirements, a special study shall be conducted.

4.1.6 The type and size of hydraulic structures should be determined through comprehensive comparison considering the cut-fill balance.

4.2 Dam Crest

4.2.1 The dam crest width shall be determined according to the requirements for structure, construction, operation, seismic resistance, etc., which should be 5 m to 12 m. The crest width of CFRD built in a highly seismic region should take the larger value. For extra-high dams, the dam crest should be wider as appropriate.

4.2.2 The freeboard shall comply with the current sector standard NB/T 10872, *Code for Design of Roller-Compacted Embankment Dams*. A special study of freeboard shall be conducted if surge might occur in the reservoir due to massive bank collapse or landslide.

4.2.3 A concrete parapet wall shall be provided on the upstream side of dam crest. The height of parapet wall should be less than 6 m, and the wall top should be 1.0 m to 1.2 m higher than the dam crest. A guardrail or curb shall be provided on the downstream side of dam crest.

4.2.4 The parapet wall for a low dam may be integrated with the face slab.

4.2.5 The elevation of the joint between parapet wall and face slab shall be higher than the normal pool level, and should not be lower than the maximum still water level under normal service conditions.

4.2.6 The camber of dam crest shall be determined based on the estimated settlement or through engineering analogies.

4.2.7 The dam portion above the elevation of parapet wall base should be placed with transition materials, and the crest road should be paved. When a traffic road is arranged on the crest, the road shall be designed in accordance with the relevant road standards.

4.2.8 A sidewalk with a width not less than 0.8 m should be provided on the upstream side of the parapet wall.

4.2.9 The parapet wall shall be subjected to stability and strength check. Structural joints shall be set in the parapet wall and be wide enough to accommodate the deformation of dam crest.

4.2.10 The dam crest shall be provided with lighting and drainage facilities. The dam crest shall be economical, practical, and aesthetic.

4.3 Dam Slope

4.3.1 The dam slopes may be determined by factors such as dam height, dam grade, filling materials, dam foundation characteristics, loads, construction and operation conditions, etc. with engineering analogies to existing dams. The determination of dam slopes should meet the following requirements:

1. When the embankment is built with hard rocks, its upstream and downstream slopes may be 1 : 1.3 to 1 : 1.4 ($V:H$); and when the embankment is built with quality natural sand and gravel materials, the slopes may be 1 : 1.5 to 1 : 1.6.

2. For extra-high dams, or dams with a design seismic intensity of Ⅷ or above, built with soft rocks, or resting on soft foundation or on the foundation with soft interlayers, the slopes should be flattened properly.

4.3.2 For Grade 1 dams, high dams and other dams with complex conditions, in addition to meeting the requirements of Article 4.3.1 of this code, the dam slopes shall be reasonably determined through stability analysis.

4.3.3 When a road is arranged on the downstream slope, the slope between two adjacent road sections may be locally adjusted, but the average slope shall meet the requirements of Articles 4.3.1 and 4.3.2 of this code.

4.3.4 The downstream slope protection may use dry boulder masonry, concrete grid frame, etc. Other types of slope protection may be adopted considering ecological environment and aesthetic requirements. The dam slope surface shall be neat, and shall have a good appearance.

4.4 Dam Zoning

4.4.1 An embankment shall be zoned according to the topographical conditions of the river valley, sources and engineering properties of materials, dam height, construction convenience and economy, and the dam zoning shall meet the following requirements:

1. Dam zoning shall facilitate the utilization of excavated materials and available quarries near the dam area.

2. The permeability of various zones should increase from upstream to downstream, and the filling materials of downstream dry zone may be exempted from this restriction.

3. The dam zoning shall be conducive to dam deformation control.

4. The angle of the interface between upstream and downstream rockfill zones with respect to plumb line may be determined by filling material properties and dam height. The dividing line between upstream and downstream rockfill zones of a high dam shall incline downstream.

5. A fine cushion zone compacted in thin layers shall be provided downstream of perimeter joint.

4.4.2 A high dam of Grade 1 or 2 shall be zoned through techno-economic comparison based on filling material test results. The zoning of an extra-high dam, or high dam built in a narrow valley or with soft rocks shall be demonstrated according to the rheology and wetting properties of filling materials and long-term deformation analysis. Other dams may be zoned with engineering analogies to existing projects.

4.4.3 Dams resting on rock foundation may be zoned as shown in Figure 4.4.3, i.e., from upstream to downstream, the cushion zone, transition zone, upstream rockfill zone, downstream rockfill zone, and downstream drainage zone.

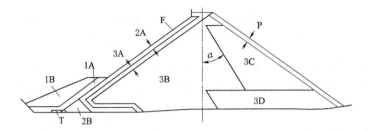

Key

1A	upstream blanket zone
1B	weighted cover zone
2A	cushion zone
2B	fine cushion zone
3A	transition zone
3B	upstream rockfill zone
3C	downstream rockfill zone
3D	downstream drainage zone
F	concrete face slab
T	plinth
P	downstream slope protection
α	angle of the interface between upstream and downstream rockfill zones with respect to the plumb line.

Figure 4.4.3　Zoning of CFRD on rock foundation

4.4.4　A concrete retaining wall may be provided at the downstream dam toe according to the downstream topographical conditions or the requirements for dam seepage monitoring, and the retaining wall shall be subjected to stability and stress calculations.

4.4.5　A transition or modulus-increased zone may be arranged on the bank slopes or behind the toe wall at the dam bottom for a narrow valley, steep bank slopes, and other unfavorable topographical conditions.

4.4.6　The cushion zone arrangement shall be determined according to the dam height, topography, geology, construction technology, and economy, and shall meet the following requirements:

1　The horizontal width of cushion zone should not be less than 3 m in the case of mechanized construction. When special measures are taken to spread materials, the width of cushion zone may be properly reduced, but the seepage stability requirements shall be met, and the transition zone shall be widened accordingly.

2　The cushion zone shall appropriately extend downstream along the contact face with bedrock, and the extending length should be determined depending on the bedrock properties, bank slope topography, and dam height.

4.4.7 When rocks are used in the upstream rockfill zone, a transition zone shall be provided between the upstream rockfill and cushion zones. The transition zone shall meet the requirements of filter and transition. The horizontal width of transition zone shall not be less than that of cushion zone, nor less than 3 m.

4.4.8 When soft rocks are used in the upstream rockfill zone of a low or medium dam, and its permeability cannot meet the drainage requirements, a chimney drainage zone shall be provided in the upstream part of the dam, and a horizontal drainage zone shall be provided along the dam base. The drainage capacity shall meet free draining requirements. In addition, a filter layer may be provided upstream of the chimney drainage zone when necessary. The rockfill materials placed in the drainage zone shall be hard and highly resistant to weathering.

4.4.9 The zoning of an embankment constructed of sand and gravel shall meet the following requirements:

1 Reliable chimney and horizontal drainage zones shall be provided and the filter requirements shall be met.

2 The top of chimney drainage zone should be higher than the still water levels under normal service conditions of reservoir, and the drainage capacity shall meet free draining requirements.

3 A transition zone may be set between the cushion zone and the sand and gravel rockfill zone if necessary.

4 A rockfill zone should be set near the downstream slope, or the slope protection should be reinforced.

4.4.10 A filter layer shall be provided on the dam foundation surface when the overburden or unfavorable geological bodies cannot meet the requirements of seepage stability. A weighted cover may be provided on the downstream dam toe when the overburden or soft interlayer in the foundation cannot satisfy the requirements of sliding stability.

4.4.11 For high dams, upstream blanket and weighted cover zones shall be provided upstream of the lower part of face slab.

4.4.12 The top elevation of the upstream blanket zone should be determined according to the dam height, sedimentation, and emptying conditions. The width of upstream blanket zone shall be determined by the material source and construction requirements. The upstream weighted cover zone shall be stable.

5 Filling Materials and Filling Criteria

5.1 Filling Materials Investigation, Testing and Quarry Planning

5.1.1 The reserve, quality, distribution and exploiting conditions of filling materials shall be investigated in accordance with the current sector standard NB/T 10235, *Specification for Investigation of Natural Construction Materials for Hydropower Projects*. When the excavated materials from project structures are used, investigation and testing of materials shall be performed as per the quarry requirements.

5.1.2 The physical and mechanical properties test of filling materials shall comply with the current standards of China GB/T 50123, *Standard for Geotechnical Testing Method*; DL/T 5368, *Code for Rock Tests of Hydroelectric and Water Conservancy Engineering*; DL/T 5355, *Code for Soil Tests for Hydropower and Water Conservancy Engineering*; DL/T 5356, *Code for Coarse-Grained Soil Tests for Hydropower and Water Conservancy Engineering*; and DL/T 5150, *Test Code for Hydraulic Concrete*.

5.1.3 The laboratory test items for filling materials shall cover specific gravity, density, absorption rate, compressive strength, softening coefficient, etc. The mineral composition test and chemical analysis should be conducted for filling materials of high dams.

5.1.4 The laboratory tests of filling materials for Grade 1 and Grade 2 high dams shall cover the following items:

1 Particle size analysis, triaxial, and compression tests.

2 Permeability test and permeability deformation test of cushion materials, sand and gravel materials, and soft rock materials.

3 Relative density test of sand and gravel materials and cushion materials.

5.1.5 The research on rheology and wetting properties of filling materials of Grade 1 high dams should be made, and the rheology and wetting tests of filling materials of extra-high dams shall be conducted. Blasting test and field compaction test shall be performed considering the site conditions for Grade 1 high dams. Additional in-situ mechanical and relative density tests shall be conducted for extra-high dams.

5.1.6 The filling material tests should be conducted on representative materials, considering the quarry characteristics, filling material properties, and construction planning for exploitation and stockpiling methods.

5.1.7 The planning for selection, exploitation, and filling of materials excavated from the project structures and quarries shall be based on the project layout and material quality requirements. Suitable excavation, transportation, processing, and stockpiling methods shall be selected. Meanwhile, the stockpiling and spoil areas shall be planned and corresponding environmental protection and water and soil conservation shall be designed. The current sector standards NB/T 10238, *Code for Design of Natural Material Source Selection and Exploitation for Hydropower Projects*; and NB/T 35062, *Design Code for Construction Planning of Rolled Earth-Rock Dam* shall be complied with.

5.2 Cushion Materials and Transition Materials

5.2.1 The cushion materials shall be well graded, meet the requirements for internal stability or self-filtration, and provide filter protection for cohesionless fill or blanket materials. The following requirements shall be met:

1. The maximum particle size should be 80 mm to 100 mm, the particles finer than 5 mm should be 35 % to 55 %, and the particles finer than 0.075 mm should be 4 % to 8 %.

2. The compacted cushion materials shall have low compressibility, high shear strength, and good constructability.

3. The permeability coefficient should be 1×10^{-3} cm/s to 1×10^{-4} cm/s, and should be 1×10^{-2} cm/s to 1×10^{-3} cm/s in cold region or for pumped storage power station.

5.2.2 The cushion materials may be crushed rock or sand and gravel or their combination. Crushed rocks shall be sourced from hard rocks that are highly resistant to weathering.

5.2.3 Fine cushion materials shall be well graded with stable internal structure, and provide filter protection for cohesionless fill or blanket materials. The maximum particle size of fine cushion zone should not exceed 40 mm, the particles finer than 5 mm should be 45 % to 60 %, and the particles finer than 0.075 mm should be 8 % to 12 %.

5.2.4 The transition materials shall provide filter and transition for the cushion materials, and shall meet the following requirements:

1. The transition materials shall be well graded, the maximum particle size should not exceed 300 mm, and the particles finer than 0.075 mm should not exceed 5 %.

2. The compacted transition materials shall have low compressibility, high shear strength, and free drainage.

3   The transition materials shall be hard rocks, which may be sourced from tunnel excavation or quarried rocks or sand and gravel.

5.2.5   When extruded curb or turnover formworks mortar is used to reinforce the upstream slope of the cushion, the compressive strength and elastic modulus of concrete or mortar at 28 days should not be greater than 5 MPa and 8,000 MPa, respectively.

5.2.6   The upstream blanket zone should be constructed of silt, silty sand, fly ash, etc. Whereas, the upstream weighted cover zone may be constructed of soil and rock spoils.

5.3   Rockfill Materials

5.3.1   For upstream rockfill zone, the compacted hard rock should be: well graded with low compressibility, high shear strength, and free drainage, the maximum particle size shall not exceed the lift thickness, the particles finer than 5 mm should not exceed 20 %, and the particles finer than 0.075 mm should not exceed 5 %. For downstream rockfill zone, the requirements of rockfill materials may be relaxed properly.

5.3.2   The upstream and downstream rockfill zones may be constructed of rocks, sand and gravel excavated from the project structures, underground caverns, or quarries, and the zoning shall meet the requirements of Section 4.4 of this code.

5.3.3   The drainage zone shall employ hard rocks that exhibit high resistance to weathering and yield a high softening coefficient. The particles finer than 5 mm shall not exceed 20 % and the particles finer than 0.075 mm shall not exceed 5 %. The compacted rockfill materials shall allow for free draining.

5.3.4   Soft rocks may be used in the dry area downstream the axis of a high dam. Following compaction, its deformation shall coordinate with that of upstream rockfill zone. In addition, soft rocks may be used in the upstream rockfill zone of a medium or low dam, but shall meet the requirements of Article 4.4.8 of this code.

5.3.5   Soft rocks should not be used in extra-high dams. The saturated compressive strength of upstream and downstream rockfill materials should not be less than 40 MPa and 30 MPa, respectively.

5.3.6   When sand and gravel are used for embankment, the zoning shall meet the requirements of Article 4.4.9 of this code. When the particles finer than 0.075 mm are more than 8 %, the materials should be used in the dry zones of the dam.

5.4 Filling Criteria

5.4.1 The filling criteria for cushion materials, transition materials, upstream and downstream rockfill materials shall be determined according to the grade and height of the dam, valley shape, seismic intensity, material properties, and the engineering analogies of existing rockfill dams with similar rock properties.

5.4.2 The porosity or relative density of rockfill materials may be preliminarily determined according to the laboratory tests and engineering experience, and shall meet the following requirements:

1. The porosity or relative density of materials in various zones should be in accordance with Table 5.4.2.

2. The filling criteria for fine cushion zone shall not be inferior to those for cushion zone.

3. The filling criteria for upstream blanket zone and weighted cover zone should be proposed according to the material properties and structural functions.

Table 5.4.2 Porosity and relative density of materials in various zones

Filling material		Cushion material	Transition material	Upstream rockfill		Downstream rockfill		Sand and gravel
				Hard rocks	Soft rocks	Hard rocks	Soft rocks	
Medium dam, low dam	Porosity (%)	15 - 20	16 - 22	20 - 24	18 - 22	20 - 25	18 - 23	–
	Relative density	–	–	–	–	–	–	0.75 - 0.85
High dam	Porosity (%)	15 - 19	16 - 21	17 - 22	17 - 20	18 - 23	17 - 20	–
	Relative density	–	–	–	–	–	–	0.85 - 0.90
Extra-high dam	Porosity (%)	15 - 18	16 - 19	17 - 20	–	18 - 21	–	–
	Relative density	–	–	–	–	–	–	0.85 - 1.00

5.4.3 The filling criteria shall be verified and modified through test fill. Both the compaction parameters and the porosity or relative density shall be used for quality control during construction, whereas the compaction parameters should be dominant.

5.4.4 Wetting requirements shall be proposed according to the rockfill material properties, and the water quantity may be determined by experience or test.

5.4.5 For the filling materials with special properties or beyond current experience, the filling criteria shall be demonstrated.

5.4.6 The requirements for porosity or relative density of modulus-increased zone shall be higher than that of upstream rockfill zone, which shall be determined through study.

6 Plinth

6.1 Alignment and Layout of Plinth

6.1.1 For rock foundation, the plinth should rest on hard and erosion-resistant bedrock that is slightly-weathered to fresh, and slightly relieved or non-relieved. For the plinth resting on completely or highly-weathered bedrock or highly relieved bedrock or the bedrock with geological defects, special treatment measures shall be taken.

6.1.2 The plinth of a medium or low dam may rest on dense overburden. The plinth of a high dam, if resting on overburden, shall be demonstrated.

6.1.3 The layout of plinth resting on rock foundation shall be determined considering topographical and geological conditions and a flat plinth should be adopted. When the abutment slope is steep, narrow plinth with equal width plus plinth extensions, inclined plinth or other types may be adopted.

6.1.4 If the topographical or geological conditions are unfavorable, plinth extension or concrete backfill may be used, and the plinth may be locally replaced by a toe wall. A special study shall be conducted when the toe wall is higher than 5 m.

6.1.5 The location of plinth should be determined by the plinth line. During excavation of the plinth foundation, the plinth line may be checked according to the geological conditions, and when necessary, re-alignment of the plinth or proper adjustment of the dam axis may be conducted.

6.2 Plinth Size

6.2.1 The width of plinth may be determined according to the allowable hydraulic gradient within the rock foundation and the foundation treatment measures, and shall meet the following requirements:

1. The allowable hydraulic gradient within the rock foundation should be in accordance with Table 6.2.1.

2. The width of plinth of a high dam should be varied in sections based on water head and the minimum width should not be less than 3 m.

3. Once the width of plinth meets the requirements for layout of grouting holes, plinth extensions may be provided to lengthen the seepage path and meet the requirements for hydraulic gradient, and filter material shall be employed to cover the plinth extensions and the downstream rock surface.

Table 6.2.1 Allowable hydraulic gradient within rock foundation of plinth

Weathering degree	Fresh, slightly weathered	Moderately weathered	Highly weathered	Completely weathered
Allowable hydraulic gradient	≥ 20	10 - 20	5 - 10	3 - 5

6.2.2 The thickness of plinth on bedrock should be approximately equal to that of the adjacent face slab, and the minimum thickness shall not be less than 0.3 m. The thickness of plinth may be varied in sections based on the water head.

6.2.3 The plinth on overburden should be connected to the cutoff wall through a concrete connection slab. Further, it shall meet the following requirements:

1. The minimum width of plinth and concrete connection slab should not be less than 3 m, and the thickness should be equal to that of the adjacent face slab.

2. The width, number, and connection mode of concrete connection slabs for a high dam or on a thick overburden shall be determined through 3D finite element analysis.

3. The concrete connection slab should be constructed following the completion of the cutoff wall and prior to first impoundment of the reservoir.

6.2.4 For a pumped storage power station, when the plinth is used for connecting the impervious elements of reservoir bottom and rim, the minimum width of the plinth should not be less than 3 m and the thickness should not be less than that of the adjacent face slab.

6.2.5 The contact face of the plinth with the face slab should be perpendicular to the face slab. The height of plinth under the face slab bottom should be 0.8 m to 1.0 m, and this requirement may be properly relaxed for the higher parts of the abutments.

6.2.6 A plinth without expansion joints may be used for the plinth on the rock foundation. In cases where abrupt changes in geology or topography are encountered, the expansion joints should be provided and staggered with the vertical joints of the face slab.

6.3 Toe Wall

6.3.1 The height of the toe wall should be determined considering topography, geological defect, and structural requirements. The top width of the toe wall shall not be less than that of the adjacent face slab. Further, the allowable hydraulic gradient of rock foundation of the toe wall should be in accordance with Table 6.3.1.

Table 6.3.1　Allowable hydraulic gradient of the toe wall rock foundation

Weathering degree	Fresh, slightly weathered	Moderately weathered	Highly weathered	Completely weathered
Allowable hydraulic gradient	≥ 20	10 - 20	5 - 10	3 - 5

6.3.2 The stability analysis shall be conducted for the toe wall. For the toe wall with a height of 5 m and above, stress and deformation analysis should also be conducted and the deformation of perimeter joints should be checked. Deep-seated sliding stability analysis shall be conducted when there are gently dipped weak discontinuities in the toe wall foundation. The load and structural calculation shall comply with the current standards of China GB/T 51394, *Standard for Load on Hydraulic Structures*; and NB/T 35026, *Design Code for Concrete Gravity Dams*.

6.3.3 The toe wall should be of the gravity type. Studies shall be performed for toe wall of other types or with special requirements.

6.3.4 Sliding stability, stress and deformation, temperature control and crack prevention of the toe wall shall comply with the current sector standard NB/T 35026, *Design Code for Concrete Gravity Dams*.

6.3.5 The Coulomb earth pressure theory should be used to calculate the earth pressure of rockfill against the toe wall, and correction should be made. When the toe wall of a high dam is 5 m or higher, the finite element method should be used to check the earth pressure of rockfill.

6.3.6 The toe wall shall meet the requirements of seepage control, and the expansion joints in the toe wall shall be staggered with the vertical joints in the face slab.

6.4　Concrete and Reinforcement of Plinth and Toe Wall

6.4.1 The requirements for plinth concrete should be the same as those for face slab concrete. The requirements for toe wall concrete shall comply with the current sector standard NB/T 35026, *Design Code for Concrete Gravity Dams*.

6.4.2 The crack control measures for plinth concrete shall be the same as those for face slab concrete. The crack control measures for toe wall concrete may comply with the current sector standard NB/T 35026, *Design Code for Concrete Gravity Dams*.

6.4.3 The single-layer two-way reinforcement should be adopted for the plinth and its extension on rock foundation. The reinforcement ratio may range from 0.3 % to 0.4 % of the flat section for each way. The two-layer two-way reinforcement should be adopted for the plinth and concrete connection slab on non-rock foundation and may range from 0.3 % to 0.4 % each layer each way. The concrete cover over reinforcement should not be less than 8 cm.

6.4.4 Toe wall reinforcement shall meet the structural and crack control requirements.

6.4.5 The plinth shall be anchored on rock foundation and the parameters of anchor bars may be determined through engineering experience. Upon encountering gently dipping discontinuities near the plinth foundation surface, the parameters of anchor bars shall be determined based on the stability requirements of the plinth and its foundation, and the resistance against grouting pressure.

7 Face Slab

7.1 Face Slab Jointing

7.1.1 The face slab shall be jointed and partitioned according to the topographical and geological conditions, dam stress and deformation analysis, construction conditions, and engineering experience. Further, the tensile and compressive vertical joints shall meet the following requirements:

1. The spacing of vertical joints should be 12 m to 16 m. For vertical joints on steep abutment or abutment with abrupt slope change, the spacing may be reduced as appropriate.

2. The vertical joints shall be perpendicular to the perimeter joints and arranged in broken line at positions approximately 1.0 m from the perimeter joint in normal direction.

7.1.2 For extra-high dams, high dams with special topographical and geological conditions, and dams with a design seismic intensity of Ⅷ and above, horizontal joints may be provided according to the results of stress and deformation analysis of face slab. The structural type of joints and waterstops shall be specifically designed, and structural measures shall be taken to prevent the dislocation of face slab.

7.1.3 The arrangement of construction joints shall consider the construction conditions, location of horizontal joints, temporary water retaining or staged impoundment requirements. The construction joints shall be perpendicular to face slab.

7.1.4 For a face slab to be placed in stages, the top of face slab at each stage shall be lower than that of the embankment. The height difference shall be determined according to the height and deformation monitoring results of the embankment and shall not be less than 5 m.

7.2 Face Slab Thickness

7.2.1 The determination of face slab thickness shall meet the following requirements:

1. The seepage hydraulic gradient shall not exceed 200.
2. The reinforcement and waterstop arrangement requirements shall be met.

7.2.2 The face slab thickness should increase gradually from top to bottom, and may be determined according to the following requirements:

1   The thickness of face slab can be calculated by the following formula:

$$t = t_0 + bH \qquad (7.2.2)$$

where

- t is the thickness of face slab (m);
- t_0 is the top thickness of face slab (m), which shall not be less than 0.3 m, and may be properly increased for extra-high dams as well as dams in severe cold regions or with a design seismic intensity of Ⅷ and above;
- b is the thickness coefficient of face slab, taken as 0.002 to 0.004;
- H is the height difference between the calculation section and the face slab top (m).

2   For low dams, the face slab may have a constant thickness of 0.3 m to 0.4 m.

7.3 Face Slab Concrete

7.3.1 The face slab concrete shall have good durability, impermeability, crack resistance, and workability, and shall comply with the current sector standards NB/T 10857, *Code for Design of Rational Service Life and Durabihity for Hydropower Projects*; DL/T 5057, *Design Specification for Hydraulic Concrete Structures*; and NB/T 35024, *Design Code for Hydraulic Structures Against Ice and Freezing Action*, and also meet the following requirements:

1   The strength class shall not be inferior to C30 for Grade 1 dams and extra-high dams, and C25 for other dams.

2   The seepage resistance shall not be lower than W12 for extra-high dams, and W8 for other dams.

3   The frost resistance shall not be lower than F100.

4   The ultimate tensile strain shall not be less than 0.85×10^{-4}, and should not be less than 1×10^{-4} for high dams.

7.3.2 The face slab concrete should use medium-heat or low-heat 42.5 MPa Portland cement, or may also use 42.5 MPa Portland cement, or ordinary Portland cement; otherwise, tests shall be conducted. The minimum consumption of cement shall comply with the current sector standard DL/T 5057, *Design Specification for Hydraulic Concrete Structures*.

7.3.3 Fly ash or other quality mineral admixtures, with certain activity and low drying shrinkage, and chemical admixtures should be mixed in the face slab concrete and shall meet the following requirements:

1. The fly ash quality shall comply with the current national standard GB/T 1596, *Fly Ash Used for Cement and Concrete*, and shall not be inferior to Grade II. The dosage of fly ash should be 15 % to 30 % of the total cementitious material. The lower and higher values should be adopted in severe cold and warm regions, respectively.

2. Air entraining admixture and superplasticizers shall be mixed in the face slab concrete. The air content shall be controlled within 4 % to 6 %. Other types of chemical admixtures that can regulate the concrete setting time may be applied if necessary.

3. The type and dosage of mineral and chemical admixtures shall be determined through tests and various chemical admixtures shall be compatible with each other.

7.3.4 The two-graded aggregate shall be used for face slab concrete. The coarse aggregate shall have a water absorption rate not exceeding 2 % and a dust content not more than 1 %. The fine aggregate shall have a water absorption rate not exceeding 3 %, and a dust content not more than 2 %; and the fineness modulus should range from 2.4 to 2.8.

7.3.5 The water-cement ratio of face slab concrete shall be less than 0.50 and 0.45 in warm and in cold and severe cold regions, respectively. The slump at the entrance of a feed chute shall meet the construction requirements and should be controlled within 3 cm to 10 cm.

7.4 Reinforcement Arrangement

7.4.1 Single- or two-layer two-way reinforcement may be adopted for the face slab. The reinforcement ratio should range from 0.3 % to 0.5 % each way each layer. For extra-high dams, two-layer two-way reinforcement should be arranged in the face slab at the top of the embankment and areas in the vicinity of the staged construction joints according to the stress and deformation analysis.

7.4.2 Anti-spalling reinforcement should be provided on both sides of the vertical compressive joints, perimeter joints, horizontal joints, and the vertical joints nearby perimeter joints for a high dam, and ties or stirrups should be arranged.

7.4.3 The thickness of concrete cover over reinforcement should not be less than 8 cm. The single-layer two-way reinforcement should be located in the middle or near upstream face.

7.5 Crack Control Measures for Face Slab

7.5.1 The base for face slab shall be smooth and free from large irregularities, local pits, and bulges.

7.5.2 An emulsified asphalt layer should be placed on the upstream face of the embankment before placing the face slab concrete.

7.5.3 The mix design of face slab concrete shall be conducted. Quality mineral and chemical admixtures shall be employed. Fibers may be applied if necessary.

7.5.4 When V-shaped grooves are set at the top of compressive joints of face slab, the depth should not be greater than 5 cm, and the bottom mortar cushion shall not occupy the effective thickness of face slab.

7.5.5 The face slab concrete should be constructed in favorable time periods and the high temperature and minus temperature time periods should be avoided. Measures shall be taken to control the placing temperature when necessary.

7.5.6 The extra fill height, pre-settlement period, and settlement rate criteria should be proposed in the concrete placement schedule for face slab. For extra-high dams, it is advisable to increase the extra fill height and extend the pre-settlement period appropriately.

7.5.7 After removal of formworks, the face slab concrete shall be timely covered for moisture retention and insulation until the reservoir impoundment or for 90 days as a minimum. In cold regions, effective insulation measures for concrete surface shall be taken till reservoir impoundment.

7.5.8 When cracks in face slab exceed 0.2 mm in width or are considered as through cracks, special treatment measures shall be taken. For dams in severe cold regions or for pumped storage power stations, more strict crack treatment criteria should apply.

7.5.9 In case cracks occur in a larger area of the face slab, auxiliary seepage control measures such as geomembrane covering may be taken following demonstration.

8 Joints and Waterstops

8.1 General Requirements

8.1.1 The waterstops shall be provided for joints of face slab, plinth, toe wall, concrete connection slab, and parapet wall to form a continuous watertight system.

8.1.2 The type of waterstops for joints shall accommodate the joint deformation.

8.1.3 The waterstop materials shall have the certificate of conformity, and be spot-checked by a certified testing agency.

8.1.4 The waterstop materials of joints shall have a good bonding property to concrete, be easy to process and install at site, and be durable.

8.1.5 The detailing of surface waterstops for face slab at and above reservoir level fluctuations in severe cold regions shall be demonstrated.

8.1.6 The use of new technologies, processes, equipment, and materials demonstrated by experiments are encouraged provided that the engineering and environmental protection requirements are met.

8.2 Perimeter Joints

8.2.1 The layout of perimeter joints shall meet the following requirements:

1. For a low dam, copper waterstops shall be set at the bottom of the perimeter joint.

2. For a medium dam, copper waterstops shall be set at the bottom of the perimeter joint and one waterstop should be set at the top of perimeter joint. The waterstop at the top may be a flexible or cohesionless filler.

3. For a high dam, in addition to the copper waterstops at the bottom of the perimeter joint, the flexible or cohesionless filler waterstops shall be set at the top of perimeter joint. Flexible filler and cohesionless filler waterstops may be jointly used. Rubber rods may be arranged in the middle of the joint.

4. For an extra-high dam, in addition to the copper waterstops at the bottom of the perimeter joint, the flexible and cohesionless filler waterstops shall be jointly used at the top of perimeter joint. Rubber rods may be arranged in the middle of the joint.

5. For a dam with special requirements, the structural type, detailing, material, and installation of perimeter joint waterstop shall be specially

studied.

8.2.2 The copper waterstop at the bottom and flexible filler at the top shall form their respective enclosed watertight systems. When there is no flexible filler at the top of vertical joint, the flexible filler at the top of perimeter joint shall be connected with the copper waterstops of vertical joint.

8.2.3 The width of perimeter joint should be 12 mm. When the vertical rib of copper waterstop is wider than the joint, the joint bottom should be widened locally. A compression-resisting board shall be set in the joint and should be fixed to the plinth.

8.2.4 A pad shall be placed under the F-type copper waterstop at the bottom of perimeter joint, which shall meet the following requirements:

1. The pad under the copper waterstops may be made of rubber, PVC, or geotextile. The thickness of pad should be 4 mm to 6 mm. A cushion shall be set under the pad. The cushion should use asphalt mortar, or may use cement mortar.

2. The gap under the pad shall be filled densely and evenly with proper fillers.

3. A rubber rod shall be placed at the top of vertical rib of copper waterstop. The rest gap of rib shall be filled with polyurethane foam boards or other flexible materials, and the rib shall be enclosed.

8.2.5 The dimensions of F-type copper waterstop shall meet the following requirements:

1. The copper waterstop tab embedded in the plinth shall not be less than 150 mm, and facilitate air vent when placing concrete. The width of the flat section on the other side of copper waterstop shall not be less than 165 mm, and height of the vertical tab embedded into the slab should be 60 mm to 80 mm.

2. The copper waterstop rib height shall be slightly greater than the possible offset of the joint, and shall not be less than 50 mm. The copper waterstop rib width should be determined according to the tangential displacement of joints and shall not be less than 12 mm.

3. The dimensioning of cross section of copper waterstop rib should comply with the current sector standard DL/T 5215, *Specification for Waterstop of Hydraulic Structure*.

8.2.6 When the flexible filler waterstop is set at the top of perimeter joint, the following requirements shall be met:

1. When the flexible filler waterstop is set at the top of perimeter joint, rubber rods or PVC rods shall be set at the surface of perimeter joint, and the rod diameter shall be greater than the expected opening of the perimeter joint.

2. The surface of flexible filler shall be provided with a protective sheet that should be fixed with expansion bolts and flat steels made of stainless steel. The protective sheets within or above the reservoir level fluctuations in severe cold regions should be fixed with countersunk head bolts and flat steel made of stainless steel.

3. If the inner side of protective sheet is coated with flexible sealing material, the protective sheet shall be firstly adhered to the concrete surface with a matching adhesive, then compressed and fixed, to ensure that both the flexible filler and the protective sheet can form an enclosed surface watertight system.

8.2.7 When cohesionless fillers are set at the top of perimeter joint, the protective cover shall be permeable and the cohesionless fillers shall not be washed out of the protective cover.

8.2.8 The vertical rib of copper waterstop shall be facing upstream.

8.3 Vertical Joints

8.3.1 According to the topographical conditions at the dam site, tensile vertical joints may be set in the face slab near the abutments, and the compressive vertical joints in the other parts of the face slab. The number and width of tensile vertical joints and compressive vertical joints shall be determined according to the topographical and geological conditions, stress and strain analysis results, and the engineering experience.

8.3.2 The layout of waterstops of vertical joints shall meet the following requirements:

1. For low dams, copper waterstops shall be provided at the bottom of tensile and compressive vertical joints.

2. For medium dams, in addition to the copper waterstop at the bottom of tensile and compressive vertical joints, a top waterstop should be provided for tensile vertical joints.

3. For high dams, in addition to the copper waterstop at the bottom of tensile and compressive vertical joints, a top waterstop shall be

provided for tensile vertical joints.

4   For extra-high dams, in addition to the copper waterstop at the bottom of tensile and compressive vertical joints, a top waterstop shall be provided for tensile vertical joints and should be provided for compressive vertical joints.

5   For dams with special requirements, the structural type, detailing, material, and construction of waterstops for vertical joints shall be specially studied.

8.3.3 For the W-type copper waterstop in tensile vertical joints, the height and width of vertical rib should be 40 mm to 60 mm, and 12 mm, respectively; the height of vertical tab should be 60 mm to 80 mm, and the respective width of two wing flats should not be less than 160 mm. In compressive vertical joints, the vertical rib height may be reduced appropriately, and should be 30 mm to 50 mm. For extra-high dams, the vertical rib height of the W-type copper waterstop of tensile vertical joints shall be specially studied.

8.3.4 The bottom of W-type copper waterstop of vertical joints shall be provided with pads, and the following requirements shall be met:

1   The W-type copper waterstop shall be set on rubber, PVC, or geotextile pad. The pad thickness should be 4 mm to 6 mm. A cement mortar layer shall be set under the pad.

2   The total width of cement mortar layer should be greater than that of copper waterstop. The strength grade of cement mortar shall not be lower than M10 and the thickness should be 5 cm to 10 cm.

3   A rubber rod shall be placed at the top of vertical rib of copper waterstop. The rest gap of rib shall be filled with polyurethane foam boards or other flexible materials, and the rib shall be enclosed.

8.3.5 When the flexible filler is set at the top of vertical joints, the protective sheet shall be provided and connected to that at the top of perimeter joint to form a protection system.

8.3.6 For high dams, or medium dams with a design seismic intensity of Ⅷ or above, the compressive vertical joint at the high-stress face slab area shall be flexible, whose width may be 10 mm to 20 mm, and a compression-resistant sheet shall be provided in the joint. The number, width, and filling material shall be specially studied for flexible vertical joints of extra-high dams. However, other vertical joint surfaces shall be coated with a thin layer of asphalt emulsion or other anti-binding materials.

8.4 Other Joints

8.4.1 The plinth on rock foundation should be provided with necessary expansion joints according to the topographical and geological conditions following excavation, which shall be staggered with the vertical joints of the face slab. The expansion joint should not be provided with a filler, and the joint face should be coated with a thin layer of asphalt emulsion or other anti-binding materials. Further, waterstop shall be set at the expansion joint, together with the waterstop of perimeter joint and bedrock, to form an enclosed watertight system.

8.4.2 When a plinth rests on overburden, the joints between plinth and connection slab and between connection slab and cutoff wall, the joints within plinth and connection slab, and the joints between connection slab and abutment shall be designed as perimeter joints with self-healing measures.

8.4.3 The construction joints may be set in the plinth according to the topographical and geological conditions, crack resistance of concrete, and construction conditions. The joint surface may be perpendicular to the plinth surface or vertical, and the plinth reinforcement shall pass through the joint.

8.4.4 In the case of the face slab connecting with the side wall of spillway or other structures, the waterstop structure of joint shall be identical to that of perimeter joints, and the selected waterstop type shall accommodate the joint deformation.

8.4.5 A flexible waterstop and protective sheet should be set at the top of the staged construction joint of face slab, and shall be connected to the surface waterstops at vertical and perimeter joints of face slab. The waterstop at permanent horizontal joints of face slab may be arranged as per the requirements for perimeter joint at the very elevation and form an enclosed system with the waterstops at vertical joints and perimeter joint. For extra-high dams and dams with a design seismic intensity of VIII or above, the location and structure of permanent horizontal joints shall be specially studied.

8.4.6 A copper waterstop shall be provided at the horizontal joint at the bottom of parapet wall, and a flexible filler waterstop shall be set at the top of joint. The waterstop at the top and bottom of joint shall be connected to the corresponding waterstop for the vertical joint of face slab.

8.4.7 A waterstop shall be set at the structural joints of parapet wall. The waterstop may be PVC, rubber, or copper waterstop and shall be connected to the copper waterstop at the horizontal joint at the bottom of parapet wall. It should be also connected to the flexible filler waterstop at the top of horizontal joint at the bottom of parapet wall.

8.5 Joint Filling and Waterstop Materials

8.5.1 Copper, stainless steel, rubber, PVC, flexible filler, and cohesionless filler may be used as waterstop materials.

8.5.2 The thickness of copper waterstop should be 0.8 mm to 1.2 mm.

8.5.3 The copper waterstop should be made of soft copper plate whose tensile strength shall not be less than 205 MPa and the elongation not less than 25 %. The test method shall comply with the current national standard GB/T 2059, *Copper and Copper Alloy Strip*.

8.5.4 In the case of larger shear displacement, the cross-section size of copper waterstop shall comply with the current sector standard DL/T 5215, *Specification for Waterstop of Hydraulic Structure*.

8.5.5 The technical specifications of stainless steel waterstop shall be in accordance with the current sector standard DL/T 5215, *Specification for Waterstop of Hydraulic Structure*, and the testing method shall be in accordance with the current national standard GB/T 3280, *Cold Rolled Stainless Steel Plate, Sheet and Strip*.

8.5.6 The thickness of PVC or rubber waterstop should be 6 mm to 12 mm.

8.5.7 Technical specifications and performance testing of PVC waterstop, rubber waterstop, and rubber rod shall be in accordance with the current sector standard DL/T 5215, *Specification for Waterstop of Hydraulic Structure*.

8.5.8 For a high dam, the flowing watertight length and penetration of flexible filler shall not be less than 150 mm and 8 mm, respectively. Other properties and testing methods shall be in accordance with the current sector standard DL/T 949, *Standard for Joint Plastic Sealant of Hydraulic Structure*.

8.5.9 In the case of design joint deformation, flexible filler shall be able to flow into the joint under water pressure and meet the waterproof requirements. The cross-sectional area of flexible filler shall be greater than that of the design joint opening.

8.5.10 The protective sheet should be 5 mm to 8 mm in thickness, and should be made of polymer waterproof materials with good aging resistance such as EPDM.

8.5.11 The homogeneous sheet or composite sheet may be selected as the protective sheet, and its technical specifications and testing shall be in accordance with the current national standard GB 18173.1, *Polymer Water-*

Proof Materials—Part 1: Water-Proof Sheet.

8.5.12 The properties of the flexible waterproof material on the inner surface of the protective sheet shall be in accordance with the current sector standard DL/T 949, *Standard for Joint Plastic Sealant of Hydraulic Structure*. The plastic waterproof material shall be well bonded to the protective sheet.

8.5.13 Fly ash or silty fine sand should be used as cohesionless filler covering the surface of joint. The maximum particle size of the cohesionless filler should not exceed 1 mm.

8.5.14 The stainless steel sheet used for protective cover of cohesionless filler should be 0.7 mm to 0.9 mm in thickness.

8.5.15 The geotextile used as the liner of the protective cover should be needle-punched nonwoven geotextile, and its technical parameters shall be in accordance with the current national standards GB/T 17638, *Geosynthetics—Synthetic Staple Fibers Needlepunched Nonwoven Geotextiles*; and GB/T 17639, *Geosynthetics—Synthetic Filament Spunbond and Needlepunched Nonwoven Geotextiles*.

8.5.16 When a rubber board is used as compression-resistant sheet, its physical properties shall meet the requirements for the J-type waterstop specified in the current national standard GB 18173.2, *Polymer Water-Proof Materials—Part 2: Waterstop*, and hot air aging shall be a mandatory testing item.

9 Foundation Treatment

9.1 Excavation of Foundation and Abutments

9.1.1 The excavated foundation surface of plinth shall be smooth and free from scarps and reverse slopes. In cases where grooves, reverse slopes or scarps impacting the compaction of cushion material exist, slope trimming or concrete backfilling shall be undertaken, or the plinth shall be relocated.

9.1.2 The foundation surface, and the upstream and downstream slopes of plinth shall meet the stability requirements. The upstream slope shall be designed as a permanent slope and shall comply with the current sector standard NB/T 10512, *Code for Slope Design of Hydropower Projects*.

9.1.3 The downstream slope above plinth foundation surface shall be flatter than face slab.

9.1.4 A rockfill embankment may be built on weathered or unloading rock foundation. The embankment foundation within the range of 0.3 to 0.5 times the dam height downstream of plinth should be of low compressibility.

9.1.5 The excavated rock bank slope of rockfill foundation within 0.3 to 0.5 times the dam height downstream of the plinth should not be steeper than 1 : 0.5. In case of a very steep bank slope, it may be excavated or backfilled with concrete to form a stable slope not steeper than 1 : 0.25, and a modulus-increased zone may be arranged at the steep slope upstream of dam axis, and its extent may be determined with reference to the stress and strain calculation results or similar projects. Scarps and overhangs that hinder the compaction of filling materials of other parts of the foundation shall be removed, or dry lean concrete or cemented masonry shall be used to smoothen the slopes.

9.1.6 The excavation and utilization of overburden may be conducted as follows:

1 The overburden is completely removed, and the plinth and embankment are built on bedrock.

2 The overburden at and within a certain range downstream of the plinth is removed, the plinth is built on bedrock and part of the embankment is built on overburden.

3 The overburden is completely or partially retained, and the plinth and embankment are built on overburden.

9.2 Treatment of Foundation Defects

9.2.1 Dam foundation treatment shall meet the requirements of seepage stability, leakage control, static and dynamic stability, allowable settlement and uneven settlement. The criteria and requirements for dam foundation treatment shall be determined according to the project-specific conditions.

9.2.2 For the foundation area of plinth and plinth extension, where geological defects such as fault fractured zones and weak intercalations exist, the impact of leakage, seepage stability, erosion, deformation and sliding on dam foundation and embankment shall be studied according to their attitude, size, and composition to determine the measures for foundation seepage treatment, seepage control, and sliding stability control. Measures such as concrete plug, cutoff wall, width increase of plinth or plinth extension, strengthening grouting and filter protection may be taken. For the geological defects in other foundation areas, corresponding measures shall be studied according to their characteristics and effects.

9.2.3 For the plinth rock foundation, consolidation grouting and curtain grouting shall be conducted, and cement grouting should be adopted. Ultrafine cement grouting or chemical grouting may be used in special cases. The requirements of cement strength grade and slurry, grouting method and grouting termination criteria for consolidation grouting and curtain grouting shall comply with the current sector standards DL/T 5148, *Technical Specification for Cement Grouting Construction of Hydraulic Structures*; and DL/T 5406, *Technical Specification of Chemical Grouting for Hydropower and Water Conservancy Projects*.

9.2.4 Consolidation grouting in the foundation shall meet the following requirements:

1. The design criteria for consolidation grouting should be determined according to the permeability, which should be 5 Lu to 10 Lu.

2. The consolidation grouting holes should be arranged in 2 to 4 rows. The row spacing may be 1.5 m to 2.0 m, and the hole spacing may be 2.0 m to 3.0 m. The normal depth shall not be less than 5 m.

3. The near-surface grouting pressure may be 0.3 MPa to 0.5 MPa, and the pressure should be increased appropriately in deep part and shall be finalized by grouting test.

9.2.5 The grout curtain should be arranged at the center of plinth and may be combined with consolidation grouting. The grout curtain should be designed according to the importance of structure, water head, geological conditions, permeability characteristics, and seepage control requirements, and shall meet the following requirements:

1   The design criteria for grout curtain, control criteria of permeability for relative impervious layer in dam foundation, depth of curtain, length of curtain in two abutments, and rows of curtain shall comply with the current sector standard NB/T 10872, *Code for Design of Roller-Compacted Embankment Dams*.

2   The row spacing and hole spacing of grout curtain shall be determined according to the geological conditions, hydro-geological conditions, water head, and grouting test data. The row spacing and hole spacing of curtain should be 1.2 m to 2.0 m. In grouting, the row spacing and hole spacing shall be adjusted according to the grouting data.

3   Curtain grouting pressure shall be determined according to geological conditions, water head, and grouting techniques. The grouting pressure shall increase gradually with the hole depth, and the termination grouting pressure for collar section should not be less than 1.0 to 1.5 times the water head. Special measures shall be taken in grouting design to improve the grouting pressure on the upper layer of foundation rock and the durability of grout curtain, and the measures shall be verified through grouting tests.

9.2.6 The quality inspection of consolidation grouting and curtain grouting shall comply with the current sector standard DL/T 5148, *Technical Specification for Cement Grouting Construction of Hydraulic Structures*, and shall meet the following requirements:

1   The quality of consolidation grouting shall be inspected by water pressure test. The number of inspection holes should not be less than 5 % of the total holes. For projects with complex geological conditions or a high dam, both water pressure and elastic wave tests of rock mass may be conducted. The number and arrangement of inspection holes and the increase of elastic wave velocity shall be determined by design.

2   The number of inspection holes in curtain grouting for water pressure test should be 10 % of the total grouting holes, and should be arranged in the parts with special conditions such as the fractured zone of foundation, large grout take, large deviation of drilling holes, and other typical formations.

9.2.7 When the plinth rests on overburden, vertical seepage control measures such as concrete cutoff wall should be employed. In addition, the filter design for seepage exit downstream of the impervious element shall be considered.

9.2.8 Design of karst treatment, concrete cutoff wall, and overburden treatment for dam foundation shall comply with the current sector standards NB/T 10872, *Code for Design of Roller-Compacted Embankment Dams*; and DL/T 5267, *Specification of Overburden Grouting for Hydropower and Water Resources Projects*.

9.2.9 Galleries for grouting, monitoring, drainage, or maintenance may be arranged in the plinth, toe wall, or dam foundation if necessary. When the curtain depth in abutments is greater than 70 m, grouting adits may be set to reduce the grouting depth of a single layer curtain.

10 Analysis and Calculation

10.1 Seepage Analysis

10.1.1 The seepage analysis of a dam shall be conducted in accordance with the current sector standard NB/T 10872, *Code for Design of Roller-Compacted Embankment Dams*.

10.1.2 For Grade 1 dams and high dams, the numerical simulation method shall be used to determine the various seepage factors of the embankment and foundation. The calculation parameters should be determined through tests and engineering analogies. Corresponding seepage analysis shall be conducted for the dam in one of the following cases:

1 Uncompleted embankment section is used to retain water in flood season.

2 Plinth rests on overburden.

3 Hanging seepage control system is used.

4 Fault, karst and fractured zone are developed.

5 There are multiple caverns near the foundation.

10.1.3 The filter criteria shall be used to check different zones of embankment and foundation, and the following requirements shall be met:

1 Cushion materials and transition materials shall meet the filter requirements.

2 The filter requirements shall be met between the upstream blanket materials and cushion materials or special cushion zone materials.

3 The filter requirements shall be met between the internal drain and surrounding rockfills.

4 When there is an erodible stratum in the dam foundation, the filter requirements shall be met between the embankment materials at the contact and the stratum.

10.1.4 At places with reverse water pressure in the embankment, cushion materials shall satisfy the seepage stability under reverse seepage, and corresponding drainage measures shall be taken if necessary.

10.1.5 For extra-high dams, seepage sensitivity analysis shall be conducted for the embankment and foundation, focusing on the leakage at the parts prone to defects or damage, considering the topographical and geological conditions, and engineering experience, to evaluate the dam seepage safety and propose the control criteria for seepage design.

10.2 Sliding Stability Analysis

10.2.1 The sliding stability analysis of a dam shall comply with the current sector standard NB/T 10872, *Code for Design of Roller-Compacted Embankment Dams*.

10.2.2 For Grade 1 dams and high dams, sliding stability calculation shall be performed for dam slopes; for other dams, corresponding sliding stability calculation shall be performed in one of the following cases:

1   Weak intercalation exists in the dam foundation or dam rests on overburden.

2   The design seismic intensity is VII or above.

3   During the construction period, the embankment is subject to overflow, or the cushion zone and priority section are used to retain water in flood season with a larger water depth.

4   The embankment is built mainly with soft rockfill or sand and gravel.

5   The topographical conditions are unfavorable.

10.2.3 For high dams, the shear strength of embankment materials shall be determined based on test results and engineering analogies. The specimen materials shall reflect the mechanical properties of embankment materials, the test method shall simulate the operating conditions of the dam, and strength should adopt nonlinear parameters. For medium or low dams, the shear strength of embankment materials may be determined through engineering analogies.

10.2.4 Stability analysis shall be conducted when the thickness of plinth or toe wall is over 2 m. The calculation requirements shall be in accordance with Sections 6.2 and 6.3 of this code.

10.2.5 Seismic stability calculation shall comply with the current sector standard NB 35047, *Code for Seismic Design of Hydraulic Structures of Hydropower Project*.

10.3 Stress and Deformation Analysis

10.3.1 The stress and deformation analysis of a dam may comply with the current sector standard NB/T 10872, *Code for Design of Roller-Compacted Embankment Dams*.

10.3.2 For Grade 1 and, Grade 2 high dams, and medium dams filled with soft rockfills or with complex foundation conditions, the 3D finite element method shall be employed to calculate the stresses and deformations.

10.3.3 Nonlinear elastic constitutive, elastic-plastic constitutive or other constitutive models that can reflect the mechanical properties of dam materials should be used in the finite element analysis of stress and deformation. For extra-high dams, two constitutive models should be used for comparative analysis. Finite element analysis shall simulate the staged loading process of embankment according to the placement and impounding process.

10.3.4 For Grade 1 high dams, and high dams built with soft rockfill, the finite element analysis of stress and deformation shall consider the effect of filling material creep deformation, predict the embankment deformation during construction and initial impound and after 5 or more years' operation, and propose the evolution of deformation and stress in typical locations.

10.3.5 The calculation parameters of filling material constitutive model required for finite element calculation of stress and deformation should be determined through tests, considering the scale effect, which may be adjusted with reference to the existing engineering experience and back analysis results. The test materials and methods shall reflect the mechanical properties of the filling materials and the construction and operation conditions of the dam.

10.3.6 In the finite element analysis of stress and deformation, the mechanical properties of the interfaces of embankment with face slab and abutment, and interfaces of face slab joints shall be reflected.

10.3.7 In the 3D finite element calculation of extra-high dams and dams with complex topographical and geological conditions, the face slab meshing shall consider the stress calculation accuracy requirements. The effect of reinforcement and ambient temperature change may be considered when necessary.

10.3.8 For extra-high dams, the parameter sensitivity analysis should be conducted considering the project features, filling material test results, and project experience. If necessary, the scale effect of parameters may be demonstrated further through numerical tests, test fill, and large laboratory tests, and the control criteria for dam stress and deformation may be proposed.

10.3.9 Dynamic stress and seismic analysis of a dam shall comply with the current sector standard NB 35047, *Code for Seismic Design of Hydraulic Structures of Hydropower Project*.

11 Seismic Measures

11.0.1 For the dam with a design seismic intensity of VII or above, under normal pool level and design flood level, the freeboard shall consider both the wave runup and crest settlement caused by earthquake, and shall meet the following requirements:

1. According to the design seismic intensity and water depth before the dam, the wave runup caused by earthquake may be taken as 0.5 m to 1.5 m.

2. The crest settlement caused by earthquake shall include the settlement of embankment and foundation, and may be determined through engineering analogies or calculations.

3. For the potential surge caused by earthquake-triggered mass collapse and landslide in the reservoir, a special study shall be performed.

11.0.2 The following seismic measures shall be taken for a dam:

1. The dam crest should be widened, or the upper part of downstream slope should be flattened with a berm at the slope change point.

2. The upper part of downstream embankment or downstream slope surface may be reinforced.

3. A lower parapet wall or an integral dam crest structure may be adopted.

4. The cushion zone may be widened to strengthen its connection to the foundation and abutments. When the abutment slope is steep, the contact between cushion zone and foundation rock may be stretched appropriately.

5. The vertical joints of face slab in the middle of riverbed may be filled with compression-resistant plates or fillers with a certain strength and flexibility.

6. The reinforcement ratio may be increased for the upper part of face slab in the middle of riverbed, particularly along the slope direction.

7. For a staged face slab, the horizontal construction joints shall be perpendicular to the face slab. Meanwhile, double-layer rebars and tie bars shall be arranged within a certain range on both sides of the construction joint.

8. Horizontal joints may be set on the middle and upper parts of the face slab in the middle of riverbed, and the joint structure and the nearby connection between face slab and the embankment shall be specially

designed.

9　Waterstop structure with a favorable deformation performance may be adopted to weaken its lessening effect on the section area of face slab.

10　The compaction criteria for rockfill may be properly raised to improve the compaction quality of the parts where the terrain changes abruptly.

11　Extra-high dams or dams built on overburden with seismic concerns shall be specially studied.

11.0.3　For a dam built with soft rock or sand and gravel, an internal drainage zone shall be set to facilitate free drain. A certain range of downstream dam slope may be built with hard rockfills, or special measures such as reinforcing the near-surface part of downstream embankment and slope surface reinforcement may be taken.

12 Staged Construction and Dam Heightening

12.1 Staged Construction

12.1.1 The placement of embankment and the staged construction of face slab shall be rationally planned based on the topography at the dam site, construction schedule, diversion, and flood protection during construction, utilization of excavated materials, reservoir impoundment, etc.

12.1.2 The staged embankment placement plan shall meet the following requirements:

1. The cushion materials and transition materials shall be placed concurrently with the neighboring rockfill, and the width of rockfill shall be greater than 20 m.

2. For a high dam, the embankment should rise uniformly, and the downstream part may be placed higher than the upstream part if necessary. However, the height difference of the parts in staged placement should not exceed 40 m.

3. The junction slope shall not be steeper than 1 : 1.3 for rockfill, and 1 : 1.5 for natural sand and gravel.

4. A temporary ramp may be set in the rockfill zone to deliver embankment materials.

5. When the uncompleted embankment is used for water retaining or overflow for flood protection during construction, the embankment zoning and staging shall meet the flood protection requirements.

12.1.3 The stages of face slab placement should be minimized. For low or medium dams, the face slab should be placed without staging. The placement timing of face slab shall avoid the peak deformation period of the rockfills underneath.

12.2 Temporary Water Retaining

12.2.1 Before placing the face slab, the priority section or embankment section may be used for water retaining in flood season, while stability against sliding and seepage shall be guaranteed.

12.2.2 The design of the embankment section for temporary water retaining shall meet the requirements for the flood protection criteria during the construction period, reservoir impoundment, emergency response, etc.

12.2.3 When the embankment is used for water retaining in flood season, the leakage shall be estimated.

12.2.4 When the embankment is used for water retaining in flood season, measures shall be taken to protect the upstream slope surface of the cushion zone.

12.3 Embankment Protection Under Overflow Condition

12.3.1 The embankment face shall be protected when overflow occurs during the construction period.

12.3.2 The allowable water-passing height of the embankment for overflow during the construction period shall be determined through techno-economic comparison. The dimensions of its sections, including plan and profile, may be determined through engineering analogies or calculations, and those for Grades 1 and 2 dams should be determined with reference to hydraulic model test.

12.3.3 The protective measures for embankment face shall accommodate the scouring on the embankment face and foundation, ensuring the embankment stability. Special attention shall be paid to protecting the connection between the embankment and abutments or downstream dam toe.

12.3.4 In the design of overflow protection, the hydraulic parameters of various operating conditions shall be calculated. The overflow protection measures for important projects should be demonstrated by test. The measures may include reinforced rockfills, gabion, reinforced concrete, roller compacted concrete, cemented sand and gravel, or their combinations. When concrete or cemented sand and gravel is used for protecting downstream dam slope, embankment drainage shall be designed.

12.4 Dam Heightening

12.4.1 A dam to be heightened in stages by plan shall be designed according to its final size. The first-built part of dam body, plinth, face slab, joints and waterstops, foundation treatment, etc. shall be constructed according to the profile and requirements of the final size.

12.4.2 For the heightening design of an existing dam, the data on the design, construction, and monitoring of the dam shall be collected and analyzed. Moreover, the existing dam shall be explored and tested to evaluate its quality and safety.

12.4.3 An existing dam may be heightened from the downstream side, and the heightening design shall meet the following requirements:

    1   Demonstrate the adaptability of existing foundation seepage control

facilities plinth, face slab, and waterstop system to the heightening.

2   Perform a dedicated design of the connection and waterstop between existing and new plinths/face slabs to form a complete seepage control system. Meanwhile, the embedded waterstops in the concrete joints between the existing and new face slabs, or plinths shall be checked and protected.

3   Study the treatment measures for the contact between the existing dam and the newly-built downstream embankment based on their material properties.

4   Analyze the stability, stress and deformation of the heightened dam.

12.4.4 The heightening design of an existing rockfill dam with an earth impervious element shall meet the following requirements:

1   Demonstrate the adaptability of existing seepage control facilities of the foundation and dam to the heightening, and reinforce them if necessary.

2   Perform a dedicated design of the connection and waterstop between existing earth impervious element and new face slab.

3   Analyze the stability, stress and deformation of the heightened dam.

12.4.5 The heightening design of an existing concrete dam shall meet the following requirements:

1   Select the connecting point between new face slab and existing dam, and check the stability, stress, and deformation of the existing dam under various loading conditions as per Section 6.3 of this code.

2   The joint between existing dam and new face slab shall be designed as a perimeter joint, and measures shall be taken to minimize joint displacement.

12.4.6 If the reservoir is required to continue operating during heightening construction, the operating level shall be studied and determined on the premise of ensuring the safety of heightening construction. The pool level shall be monitored during construction, and relevant operating requirements shall be met.

13 Safety Monitoring

13.0.1 The safety monitoring design of a dam shall comply with the current sector standard DL/T 5259, *Technical Specification for Earth-Rockfill Dam Safety Monitoring*. The monitoring shall commence in time and the initial data shall be obtained after installation of instruments. The monitoring data shall be complied and analyzed systematically and periodically in accordance with the current sector standard DL/T 5256, *Data Compilation Code for Earth-Rockfill Dam Safety Monitoring*.

13.0.2 The layout of the instruments shall observe the following principles:

1. Reflect the working status of the dam.
2. Combine the permanent monitoring with temporary monitoring during the construction period.
3. Coordinate the internal and external monitoring, and combine the cable laying and protection of internal and external monitoring instruments.
4. Arrange at least one monitoring section at the cross section with the maximum dam height, and may set monitoring sections at other cross sections or longitudinal section along dam axis if necessary.
5. Minimize the interference between instrument installation and construction and facilitate monitoring.

13.0.3 For high dams, the following items shall be monitored, and the number of monitoring items may be reduced as appropriate for low and medium dams:

1. Vertical and horizontal displacements of dam surface.
2. Vertical displacement in the dam and horizontal displacement of the dam in flow direction.
3. Displacement of joints.
4. Deformation, strain, and temperature of face slab.
5. Stresses of reinforcement of face slab.
6. Leakage and seepage pressure of dam and foundation.
7. Seepage around the dam.

13.0.4 The following monitoring items may be added if necessary:

1. Earth pressure of rockfills.
2. Seismic response.

3 Horizontal displacement of dam, normal to flow direction.

4 Void below face slab.

5 Deformation of the contact between dam and foundation.

6 Foundation settlement.

7 Thrust force of ice cover on face slab in cold regions.

8 Leakage zoning.

9 Water quality of leakage through dam and foundation.

13.0.5 The monitoring of the dam on overburden shall meet the following requirements:

1 For a foundation on thick overburden, monitoring points for deformation and seepage pressure shall be arranged on the surface of and inside the foundation, and on the diaphragm wall to monitor the settlement of overburden, seepage pressure and its distribution and change, and the deformation and stress of diaphragm wall.

2 For Grade 1 dams or Grade 2 high dams, a cutoff wall shall be set at the bottom of measuring weir. When unavailable, seepage pressure monitoring instruments may be arranged behind the dam to estimate the leakage.

13.0.6 For the toe wall with a height of 5 m or above, the perimeter joints shall be subjected to 3D displacement monitoring. Meanwhile, the earth pressure acting on the toe wall, pressure acting on the toe wall bottom, and seepage pressure should be monitored as well.

13.0.7 For high dams or dams built in a narrow valley, displacement meters should be arranged at places where the dam connects to steep abutments or abutments with abrupt changes to monitor the contact displacement of dam and abutments.

13.0.8 For extra-high dams, the horizontal displacement of dam in the direction normal to flow shall be monitored. For key parts in the dam, instruments of different types should be arranged with a certain redundancy.

13.0.9 Monitoring points shall be arranged on the upstream slope to monitor the surface deformation in construction.

13.0.10 The internal dam deformation may be monitored continuously and automatically using a flexible inclinometer, and a pipe robot may be used when conditions permit. Once demonstrated, monitoring galleries may be arranged at different elevations within the dam. The instruments may be placed in the galleries directly or placed in holes drilled on the galleries for internal monitoring. The instrument cables shall be led to galleries and be protected.

13.0.11 The deflection of face slab may be monitored continuously and automatically using the flexible inclinometer. The deformation of dam surface may be monitored by Global Navigation Satellite System (GNSS), Interferometric Synthetic Aperture Radar, or fiber-optic gyroscope.

13.0.12 The expected range of monitoring values and the type of instruments shall be determined based on the calculation results with reference to the actual monitoring results of similar projects. The selection of instruments shall follow the principles of reliability, durability, economy and practicability, and automatic monitoring is encouraged. The instruments in cold regions shall be free of freezing.

14 Construction Requirements

14.0.1 In construction stage, the technical requirements for excavation and support of dam foundation, foundation treatment, embankment staging, embankment filling, concrete placement, and construction of joints and waterstops shall be specified. Construction requirements and productive test requirements such as dam material source check, foundation grouting, embankment placement and compaction, and concrete mix proportion should be proposed according to the project characteristics.

14.0.2 According to previous study, the principle of material source utilization shall be clearly defined at the construction stage, and the requirements of material source review shall be proposed to check the quality and reserve of material. In different construction periods, the cut-fill balance shall be checked according to the topography and geology, access road, and excavation condition of the quarry. The exploitation plan shall be adjusted dynamically according to the material balance, and the dam zoning may be adjusted if necessary.

14.0.3 Prior to construction, the placement intensity, stockpiling plan, construction schedule, subsequent diversion scheme, and reservoir impounding plan shall be checked according to the quarrying intensity, transport capacity, and actual construction progress.

14.0.4 The dam zoning and design parameters of dam material shall be adjusted dynamically according to material source check, exploitation plan, and the test fill, and the gradation and filling criteria of dam materials shall be checked according to construction inspection results.

14.0.5 During the construction period, the design of dam foundation excavation, plinth foundation surface, and grouting parameters shall be checked according to the revealed geological conditions and grouting test results.

14.0.6 A drainage for reverse seepage shall be designed when the elevation of plinth foundation surface is lower than that of rockfill foundation surface or where the downstream water level might be higher than plinth foundation surface in construction.

14.0.7 During the construction period, the requirements for face slab concrete mix proportion, temperature control and crack prevention shall be proposed, and the protection measures for the waterstops at perimeter joint at the construction stage shall be proposed according to the previous test results or engineering analogies.

14.0.8 During the construction period, the slope surface of upstream cushion zone may be fixed by extruded curb, turnover formworks mortar, or protected by compacting low strength mortar, spraying emulsified asphalt, shotcrete, etc., and the evenness requirements shall be proposed.

14.0.9 According to the requirements for construction planning and dam deformation control, the principle of embankment zoning and staging, face slab staging, and the flood protection scheme shall be proposed. The control criteria for the height of the embankment over face slab top, pre-settlement period, and deformation rate of upstream slope prior to face slab concreting, as well as the construction timing of dam crest structure, shall be specified. Measures for accelerating wetting deformation and reducing subsequent deformation for embankment, such as water filling preloading, may be proposed if necessary.

14.0.10 Construction quality control criteria and testing requirements shall be proposed at the construction stage. For high dams, the requirements for real-time monitoring of construction quality, information management, and smart construction should be proposed.

14.0.11 Requirements for installation and protection of safety monitoring facilities, monitoring time and frequency, monitoring data recording, and compilation shall be proposed at the construction stage.

14.0.12 If abnormalities occur during construction, the causes shall be analyzed based on inspection and monitoring data, and treatment measures shall be proposed. When defects are found during construction, the requirements for inspection and treatment shall be proposed.

14.0.13 For Grade 1 high dams, back analysis shall be conducted timely based on the actual construction process, quality inspection results, and monitoring data.

15 Initial Impoundment, Operation and Maintenance

15.0.1 Prior to the initial impoundment, physical progress and plugging requirements shall be proposed according to the impoundment plan, including:

1 The requirements for physical progress of dam and seepage control works, water release structures, energy dissipation, and erosion control structures, as well as resettlement and related facilities.

2 The plugging requirements for construction adits, access tunnels, exploratory adits, and exploratory boreholes that affect the impoundment.

15.0.2 During the impoundment period, the limits of rising and descending rates of pool level shall be proposed according to the project characteristics. For high dams and large reservoirs, the water level control requirements of staged impoundment shall be proposed according to the plugging and impoundment plan.

15.0.3 During the initial impoundment period, the requirements for operation and safety monitoring, patrol inspection, defect treatment and risk prevention and control of the dam and related structures shall be specified according to the project characteristics and impoundment plan, including:

1 Operation requirements for dam, water release structures, energy dissipation, and erosion control structures.

2 Technical requirements for safety monitoring of dam, water release structures, energy dissipation, and erosion control structures and bank slopes near the dam.

3 Technical requirements for subsequent defect treatment.

4 Principles for prevention and control of main risks and emergency handling of abnormalities.

15.0.4 During the initial impoundment period, feedback analysis and evaluation of dam safety should be conducted according to impoundment process and monitoring results for extra-high dams and high dams with complex conditions.

15.0.5 During the operation and maintenance (O&M) period, the requirements for O&M management shall be proposed according to the project characteristics and the reservoir operation scheme, including:

1 Requirements for routine patrol inspection and O&M management of

dam.

2 Technical requirements for maintenance of safety monitoring facilities of dam and safety observation.

3 Requirements for safety analysis and evaluation of dam.

4 Requirements for O&M management of water release structures.

5 Requirements for main risk prevention and control and emergency response plan.

15.0.6 When the seepage pressure, or the quantity or quality of leakage water is found abnormal, the underwater face slab, plinth and waterstops shall be systematically analyzed or inspected underwater in time, and the reservoir may be emptied for inspection if necessary. For extra-high dams, inspection, analysis, and evaluation of face slabs and joint waterstop system should be conducted regularly.

15.0.7 When encountering extreme hydrometeorological conditions, intensive earthquake, etc., dam and water release structures shall be inspected thoroughly, and the design basis of original hydrometeorology and project adaptability shall be checked and evaluated.

15.0.8 During the O&M period, safety evaluation of dam operation shall be conducted regularly according to the safety monitoring data and patrol inspection results. Back analysis should be conducted for extra-high dams to predict and evaluate the late-stage performance of the dam. If the monitoring data is abnormal or exceeds the design allowable value, causes shall be identified in time and necessary treatment measures shall be proposed.

15.0.9 During the O&M period, when major defects and safety risks regarding dams are found, the causes shall be analyzed and the safety risk shall be assessed in time. When necessary, special inspection, detection, or investigation may be conducted and a special treatment plan may be proposed.

Explanation of Wording in This Code

1. Words used for different degrees of strictness are explained as follows in order to mark the differences in executing the requirements in this code.

    1) Words denoting a very strict or mandatory requirement:

        "Must" is used for affirmation; "must not" for negation.

    2) Words denoting a strict requirement under normal conditions:

        "Shall" is used for affirmation; "shall not" for negation.

    3) Words denoting a permission of a slight choice or an indication of the most suitable choice when conditions permit:

        "Should" is used for affirmation; "should not" for negation.

    4) "May" is used to express the option available, sometimes with the conditional permit.

2. "Shall meet the requirements of…" or "shall comply with…" is used in this code to indicate that it is necessary to comply with the requirements stipulated in other relative standards and codes.

List of Quoted Standards

GB/T 50123,	*Standard for Geotechnical Testing Method*
GB 50199,	*Unified Standard for Reliability Design of Hydraulic Engineering Structures*
GB 50201,	*Standard for Flood Control*
GB/T 51394,	*Standard for Load on Hydraulic Structures*
GB/T 1596,	*Fly Ash Used for Cement and Concrete*
GB/T 2059,	*Copper and Copper Alloy Strip*
GB/T 3280,	*Cold Rolled Stainless Steel Plate, Sheet and Strip*
GB/T 17638,	*Geosynthetics—Synthetic Staple Fibers Needlepunched Nonwoven Geotextiles*
GB/T 17639,	*Geosynthetics—Synthetic Filament Spunbond and Needlepunched Nonwoven Geotextiles*
GB 18173.1,	*Polymer Water-Proof Materials—Part 1: Water-Proof Sheet*
GB 18173.2,	*Polymer Water-Proof Materials—Part 2: Waterstop*
NB/T 10235,	*Specification for Investigation of Natural Construction Materials for Hydropower Projects*
NB/T 10238,	*Code for Design of Natural Material Source Selection and Exploitation for Hydropower Projects*
NB/T 10512,	*Code for Slope Design of Hydropower Projects*
NB/T 10857,	*Code for Design of Rational Service Life and Durability for Hydropower Projects*
NB/T 10872,	*Code for Design of Roller-Compacted Embankment Dams*
NB/T 35024,	*Design Code for Hydraulic Structures Against Ice and Freezing Action*
NB/T 35026,	*Design Code for Concrete Gravity Dams*
NB 35047,	*Code for Seismic Design of Hydraulic Structures of Hydropower Project*
NB/T 35062,	*Design Code for Construction Planning of Rolled Earth-Rock Dam*
DL/T 949,	*Standard for Joint Plastic Sealant of Hydraulic Structure*

DL/T 5057,	*Design Specification for Hydraulic Concrete Structures*
DL/T 5148,	*Technical Specification for Cement Grouting Construction of Hydraulic Structures*
DL/T 5150,	*Test Code for Hydraulic Concrete*
DL 5180,	*Classification and Design Safety Standard of Hydropower Projects*
DL/T 5215,	*Specification for Waterstop of Hydraulic Structure*
DL/T 5256,	*Data Compilation Code for Earth-Rockfill Dam Safety Monitoring*
DL/T 5259,	*Technical Specification for Earth-Rockfill Dam Safety Monitoring*
DL/T 5267,	*Specification of Overburden Grouting for Hydropower and Water Resources Projects*
DL/T 5355,	*Code for Soil Tests for Hydropower and Water Conservancy Engineering*
DL/T 5356,	*Code for Coarse-Grained Soil Tests for Hydropower and Water Conservancy Engineering*
DL/T 5368,	*Code for Rock Tests of Hydroelectric and Water Conservancy Engineering*
DL/T 5406,	*Technical Specification of Chemical Grouting for Hydropower and Water Conservancy Projects*